Rodinia

The Mother of Supercontinents

JOHN S. PRITCHETT

Table of Contents

Chapter 1
The World You Could Not Walk On

At first, it does not resolve into a single place.

A cluster of large landmasses fills the view (see Figure 1.1). They sit close together, their edges pressed into one another, but not in a way that forms a clean, continuous surface. The shapes do not align perfectly. The boundaries are uneven. This is not a

Figure 1.1 — Reconstruction of Rodinia (Proto-Rodinia)

~1.04 billion years ago

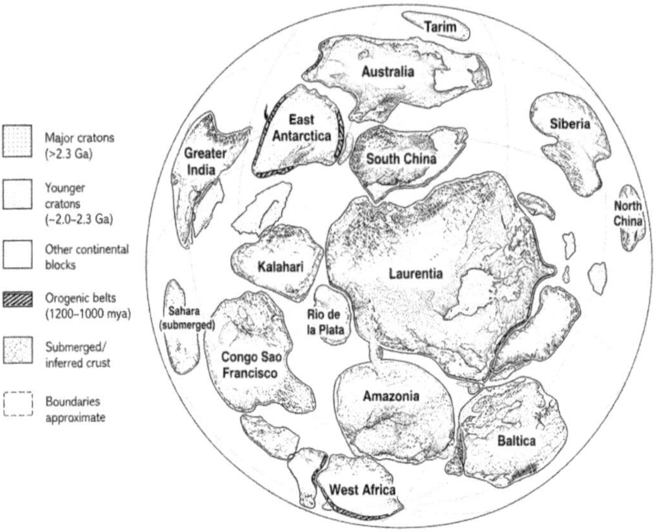

Rodinia (Proto-Rodinia) was assembled through a series of continental collisions between ~1.2 and 1.0 billion years ago. Orogenic belts mark sutures where continents joined. Boundaries and positions of some blocks remain approximate.

finished continent, but a system in the process of becoming one.

It does not feel familiar.

There is no single horizon to follow, no uninterrupted stretch of land that clearly defines itself as a continent. Instead, there are pieces, large, stable blocks of crust, held together by zones of collision. The connections are real, but they are not simple.

What appears here is a reconstruction.

These landmasses were assembled through immense forces acting over long periods of time. Ancient continental cores, including Laurentia, Baltica, Amazonia, and West Africa, were driven together. Where they met, the crust did not merge cleanly. It fractured, folded, and rose, forming long belts of mountains along the zones where the collisions occurred.

These boundaries matter more than the land itself.

Standing on this surface, you do not stand on a unified continent in the way we imagine one today. You stand on a joined system, sections of crust forced together, each with its own history, connected along lines of stress and deformation.

The land exists, but it is not yet a coherent whole.

The surface feels hard and exposed. There is no vegetation to soften the ground, no grass, no forests, no biological layer shaping the landscape. Beneath your feet is bare rock and weathered material, broken down slowly by wind, water, and temperature change. There are no developed soil systems, nothing like the layered structure that today holds water, nutrients, and roots in place.

When rain falls, it does not settle; it moves.

Without soil or plant life to retain it, water runs quickly across the surface, carrying loose material with it. Channels form and disappear. Flow patterns shift. Retention is limited. Stability is rare.

Rain falls, runs off, and leaves very little behind.

The atmosphere above does not provide the same level of protection it does today.

Oxygen is present, but not in a steady balance. It is being produced and consumed at the same time, fluctuating over long periods. The ozone layer exists, but it is not consistently stable. At times and in some regions, higher levels of ultraviolet radiation reach the surface, limiting where life can persist and how far it can spread.

There is no canopy to filter light. No cover to reduce exposure.

The danger is not sudden. It is constant.

Temperature patterns across the land are uneven and often extreme. The clustered landmasses create large interior regions that lie far from moderating ocean influence. These areas receive less moisture and experience wider swings between heating and cooling. Heat builds during the day and is lost at night. Seasonal variation is more pronounced.

The result is instability.

From a distance, the landscape appears usable. Solid ground stretches outward. Space exists. The physical platform is there.

But function is missing.

The land does not behave the way land does today.

There is no integrated soil system, no reliable retention of water, no biological layer to stabilize the surface. These are not minor differences. They are structural limitations that prevent the land from supporting complex, persistent systems.

Life already exists at this time in Earth's history. It is not absent.

But it does not establish itself across the land in a way that reshapes it. It does not form widespread ecosystems on the surface. It does not stabilize the environment or transform it into something self-sustaining.

Instead, life remains concentrated in a different part of the system.

To understand why, attention must shift away from the land and toward the oceans.

In the oceans, conditions are more stable. Water absorbs and scatters ultraviolet radiation, reducing exposure. It moderates temperature change, preventing the extremes found on land. Chemical processes occur in a more controlled environment. Nutrients dissolve, circulate, and accumulate.

The ocean acts as a buffer.

It creates conditions where life can persist, adapt, and gradually expand in complexity.

Life is not missing from this world; it is restricted.

That distinction is important.

This is not a lifeless Earth. It is a constrained one. The essential components of the planetary system, land, water, atmosphere, and energy from the Sun, are all present. But they remain out of alignment, unable to support life beyond limited environments or allow it to reshape the surface of the planet.

Rodinia earns the name "Mother of Supercontinents" not because it began everything, but because it carried Earth through a crucial stage of preparation, constraint, breakup, and renewal.

The system is close, but it is not there yet. The land stretches across the planet, available in a physical sense, but under these conditions it cannot yet be used.

Chapter 2
Protection That Does Not Hold

The surface of Rodinia is exposed for a reason.

The land is not unusable because it is too far from water or too difficult to reach. It is unusable because the systems that protect life at the surface are not working together in a consistent way.

On the modern Earth, protection from radiation comes from two main sources. One operates in the upper atmosphere. The other surrounds the planet at a much larger scale.

Both exist during the time of Rodinia.

But neither operates with the same consistency seen today.

The first layer of protection is the ozone layer.

Ozone is a form of oxygen that absorbs ultraviolet radiation from the Sun. When ozone is present in sufficient amounts, it reduces the amount of harmful radiation that reaches the surface. This allows life to exist outside of water without constant damage from ultraviolet exposure.

During the time of Rodinia, oxygen is already present in the atmosphere. Earlier in Earth's history, microorganisms had released enough oxygen for some of it to accumulate in the air. Scientists call this shift the Great Oxidation Event. That earlier increase allows an ozone layer to form and provide real protection.

But that protection is not as stable or as consistently maintained as it is today.

Oxygen is being produced by microorganisms in the oceans and consumed by chemical reactions at the same time. Iron in the oceans, volcanic gases, and exposed rock surfaces all react with oxygen. Over long periods, this changes the mix of gases in the atmosphere.

Because ozone forms from oxygen, the ozone layer likely varies as well (see Figure 2.1). At times, it provides meaningful protection. At other times, that protection weakens, allowing higher levels of ultraviolet radiation to reach the surface.

Protection exists, but it is not constant.

The second layer of protection is the magnetic field.

Figure 2.1 – Ozone Formation and Vartability

Ozone protection varies with oxygen levels.

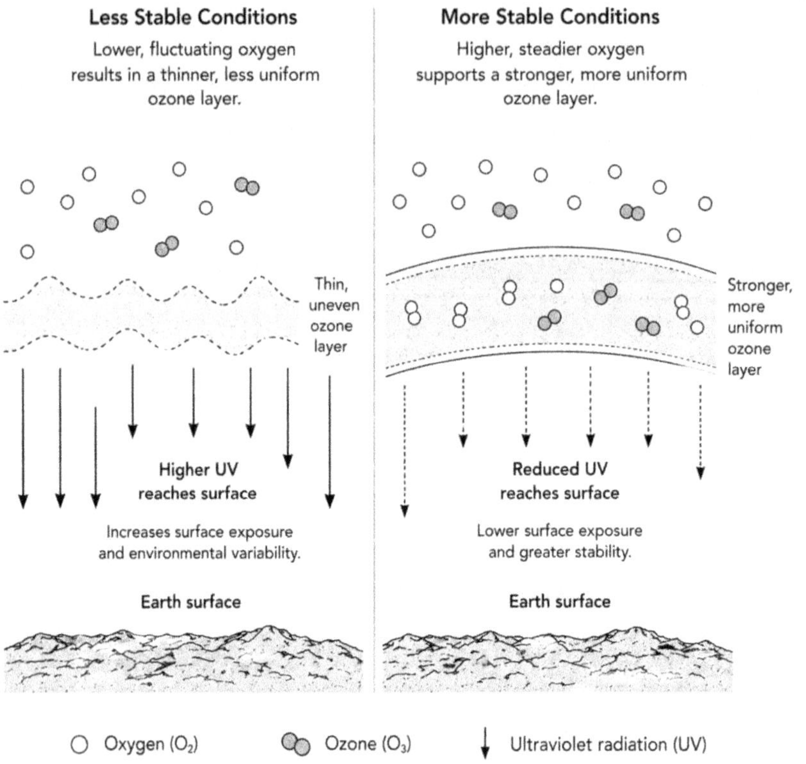

Ozone forms from oxygen in the atmosphere. When oxygen levels are low or fluctuate, the ozone layer becomes thinner and less uniform, allowing more ultraviolet radiation to reach Earth's surface. When oxygen is higher and steadier, the ozone layer strengthens, reducing ultraviolet radiation and supporting more stable surface conditions.

The Earth generates a magnetic field from motion within its core. This field extends far into space and interacts with charged particles from the Sun, known as the solar wind. When the magnetic field is strong, it deflects much of this incoming radiation before it reaches the atmosphere.

During this period in Earth's history, the magnetic field is already well established. It likely provides substantial protection, much as it does today.

However, like all planetary systems, it is not perfectly static. Its strength and structure can vary over long timescales as conditions within the core change.

These variations do not eliminate protection.

But they can influence how consistently it is applied.

These two systems are connected.

The magnetic field reduces the number of high-energy particles entering the atmosphere. The ozone layer absorbs ultraviolet radiation that reaches lower levels. Together, they form a layered system of protection.

On the modern Earth, that system is stable enough to support life across a wide range of environments on land.

During Rodinia, that layered protection likely operates, but with more variability over time.

At times, ultraviolet radiation can penetrate more deeply. At other times, interactions between solar particles and the atmosphere can influence the chemistry of the atmosphere. These changes do not make the surface uninhabitable, but they make conditions less predictable.

For simple organisms living in the oceans, this is manageable.

Water absorbs and scatters radiation. Even a few meters of depth can reduce exposure significantly. It also moderates temperature and provides a stable chemical environment.

On land, there is no such buffer.

Exposure at the surface is direct.

This is one of several factors that limits the expansion of life onto land.

The issue is not the absence of protection.

It is the variability of protection, combined with other constraints.

Life requires conditions that are not only survivable, but repeatable. It must persist across cycles of day and night, across seasons, and across longer environmental changes. When protective systems vary too much, or when they are not reinforced by other stabilizing processes, those conditions are harder to maintain.

Rodinia presents a world where the necessary systems are in place, but they still do not operate together in a fully stable and reinforcing way.

The atmosphere contains oxygen, but not at levels that consistently support a modern-style ozone layer.

The magnetic field is active, but part of a system that can vary over time.

Radiation is reduced, but not uniformly across all conditions.

The result is a surface environment that cannot be depended on in the same way as today.

This is not the failure of a single system. It is a problem of coordination: the components are present, but they remain out of alignment.

Because of this, the land remains exposed, and life remains largely confined to environments where additional protection and stability exist.

That is why the story has to move from the exposed surface to the oceans, where water provides the steadier conditions the land cannot yet offer.

Chapter 3
The Only Safe Place

If the land cannot be used, the question becomes simple.

Where does life remain?

The answer is not spread across the surface. It is concentrated in one place.

The oceans.

This is not a preference. It is a constraint.

On Rodinia, the oceans are not just one environment among many. They are the primary environment where conditions remain stable enough for life to persist over time.

Water changes the system.

The same radiation that reaches the land is reduced as it enters the ocean. Ultraviolet radiation is absorbed and scattered within the upper layers of water. Even a short distance below the surface, exposure drops significantly, often to levels that simple organisms can tolerate.

On land, exposure is direct; in water, it is reduced.

Temperature behaves differently as well.

Air heats and cools quickly. Rock responds to these changes, leading to large swings in temperature over short periods. Water absorbs heat more slowly and releases it more slowly. This moderates temperature change and creates a more consistent environment.

Consistency matters.

Life does not require perfect conditions. It requires conditions that persist long enough for processes to continue.

The oceans provide that persistence (see Figure 3.1).

Chemistry is also more continuous in water.

Elements dissolve, mix, and circulate. Nutrients can move through the system instead of remaining fixed in one place. Chemical reactions occur in a medium that allows repeated interaction over time. This continuity is critical for early life, which depends on basic chemical processes to obtain energy and build structure.

On land, these processes are limited.

Figure 3.1 — Ocean as the Primary Habitat

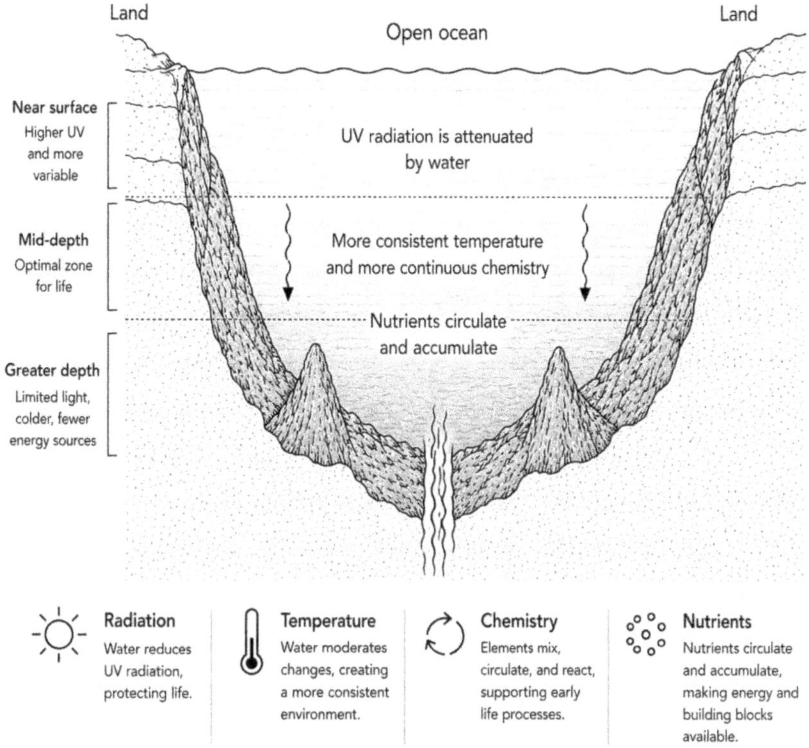

Life persists in more stable marine environments while land remains exposed and unstable.

Water moves quickly across exposed rock. There is little opportunity for sustained interaction. Materials are transported away rather than retained.

In the ocean, materials remain in contact.

Reactions can repeat.

Systems can develop.

Life takes advantage of this.

The organisms that exist during this time are simple. They are microscopic, often forming layered communities in shallow marine environments. Many of these communities build structures known as stromatolites, created as microorganisms trap and bind sediment over time.

These structures grow slowly.

They provide clear evidence that life is present and active, even within a constrained system.

Their location is not random.

They are found where conditions are most favorable. Shallow water allows access to sunlight, which some microorganisms use as an energy source.

At the same time, the water above them reduces radiation and moderates environmental change.

This creates a narrow zone where life can function.

Too deep, and light becomes limited.

Too shallow, and exposure increases.

Life occupies the middle, and that narrow position matters.

It shows that even within the oceans, conditions are not uniform. Only certain regions provide the balance required for survival. These regions act as stable zones within a larger system that remains variable.

The oceans are conditionally stable.

They reduce the most immediate stresses found on land, but they do not eliminate all challenges. Oxygen levels can still vary. Chemical conditions continue to shift. Environmental changes still occur.

But compared to the land, the oceans provide continuity.

Life does not expand across Rodinia because it cannot maintain itself outside these conditions. The systems required for broader expansion, stable protection, sustained chemistry, and biological reinforcement, are not fully in place.

It remains where the system allows it to persist.

That confinement shapes everything that follows.

The land remains exposed and largely unused.

The oceans carry the full burden of sustaining life.

All biological activity is concentrated within a limited set of environments.

Rodinia is not empty. It is uneven.

Life exists, but only within boundaries set by protection, continuity, and chemistry. Beyond those boundaries, the system does not support expansion.

The first hint of change appears where those boundaries blur, along the restless edges where land and water meet and where stability gives way to movement.

Chapter 4
The Edges Where Change Happens

The oceans provide stability. The land provides exposure. Between them lies a third environment: the boundary.

This is where conditions begin to change.

The edge of the ocean is not a fixed line. It moves continuously, shaped by tides, waves, storms, sediment, and changes in sea level. On Rodinia, this movement may be especially important. The Moon is closer to Earth than it is today, which likely contributes to stronger tidal effects in many regions.

Water advances.

Water retreats.

The boundary shifts.

In many coastal areas, shallow regions are covered and uncovered repeatedly. At one moment, they are part of the ocean. Later, they are exposed to air, sunlight, heat, and drying. Then the water returns.

This creates a different kind of environment.

Not stable like the deeper ocean.

Not fully exposed like the land.

But variable.

These tidal zones experience repeated change. When covered by water, they benefit from some of the protections found in the ocean. Radiation is reduced.

Temperature is moderated. Chemical processes can occur in a fluid environment.

When exposed, those protections are reduced.

Radiation increases.

Temperature changes more rapidly.

Water is lost.

This repeated cycle creates stress.

But it also creates opportunity.

Materials move through these regions more actively. Sediment is deposited and removed. Dissolved nutrients from the ocean interact with minerals from the land. Chemical gradients form, meaning that chemistry differs from one place to another, and those patterns form and reform as conditions shift.

These gradients matter.

They create differences in concentration, temperature, moisture, and exposure that can drive chemical reactions. In more stable environments, change tends to be gradual. In variable environments, reactions are repeatedly disrupted and re-established.

The tidal zone becomes a place where processes repeat under slightly different conditions.

This repetition is important.

Life depends on cycles. It depends on the ability to carry out the same basic processes again and again. In the open ocean, conditions are more stable, but change can be slow. On land, conditions are often too harsh for most biological processes to continue.

At the boundary, conditions fluctuate.

This fluctuation allows systems to be tested.

Water brings dissolved materials into contact. Exposure alters those materials. Re-submersion allows reactions to continue. Each cycle introduces variation.

Some variations do not persist.

Others do.

The organisms that exist during this time are still simple, but some can occupy shallow environments where conditions change more often. Microbial communities, similar to those that form stromatolite systems, can develop in these zones. They are not land organisms in the modern sense. They still depend on water. But they are no longer confined only to deeper, more stable marine settings.

These organisms do not expand across the land.

But they begin to operate at its edge.

The shift is subtle.

Life remains tied to environments where water is present, but it begins to occupy places where conditions are less predictable. It starts to exist closer to exposure, closer to drying, closer to the direct interaction between ocean, atmosphere, and rock.

The boundary becomes active (see Figure 4.1).

It is not safe or stable, but it is dynamic.

Dynamic systems allow change.

This does not mean that the land becomes usable. The same limitations still apply. Radiation remains a challenge. Water does not persist for long. There is no developed soil system to support continued growth.

But the edge introduces a new pattern.

Figure 4.1 — Continental Margins: Zones of Interaction

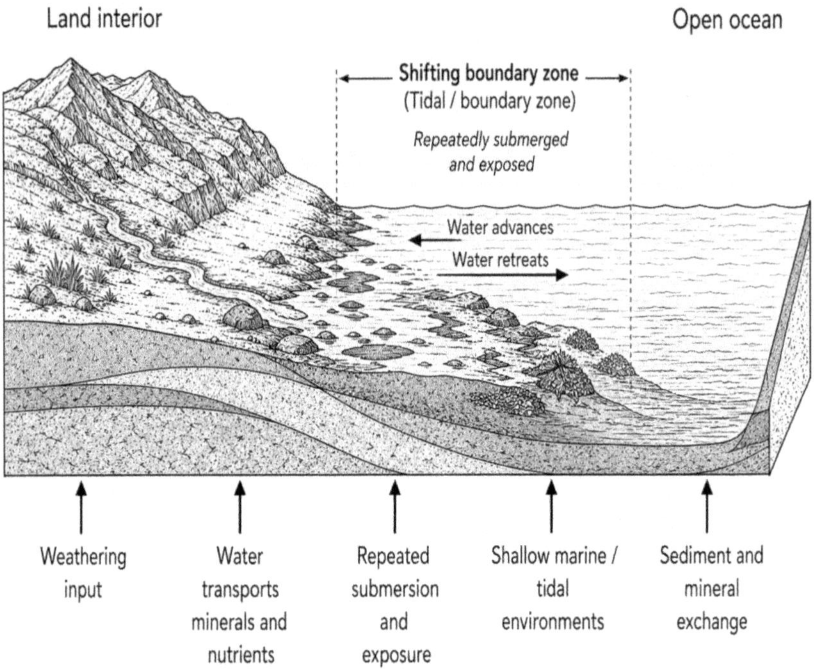

Continental margins create shifting boundary zones where water, rock, sunlight, and chemistry interact repeatedly.

Conditions vary.

Processes repeat.

Outcomes differ.

Over time, this variation becomes important.

It allows small changes to accumulate. It allows systems to respond to fluctuation rather than depend entirely on stability. It creates environments where land, water, and atmosphere interact directly.

Rodinia as a whole remains constrained. The oceans still support most life, and the land remains largely unusable, but at the edges the system begins to behave differently.

This is where new patterns of change begin.

Not in the most stable environments, and not in the most extreme ones, but in the regions where conditions shift between the two.

Those shifting margins matter because they belong to a larger planetary system, one whose climate can amplify small changes into global instability.

Chapter 5
A Planet That Overcorrects

Rodinia is not a finely balanced system.

The land is exposed.

The oceans carry life.

The boundaries shift.

But the larger problem is not any one of these conditions.

It is how the system responds to change.

On the modern Earth, climate is regulated by a balance of processes. When temperatures rise, certain mechanisms can reduce that increase. When temperatures fall, other processes can help restore warmth. These responses do not eliminate change, but they limit extremes.

During the time of Rodinia, and especially during its later breakup, that balance may have been less effective.

The system responds.

But not always in a controlled way.

At times, it overcorrects.

One of the clearest examples of this behavior is global cooling.

In many reconstructions, Rodinia places large areas of exposed land in positions where rainfall and weathering can be significant. When rain falls on exposed rock, it begins a process known as chemical weathering. Carbon dioxide in the atmosphere dissolves in rainwater and reacts with minerals in the rock. This process removes carbon dioxide from the atmosphere and eventually transfers it into the oceans.

Carbon dioxide is an important greenhouse gas. It helps retain heat in the atmosphere.

When carbon dioxide is removed, the atmosphere retains less heat.

Temperatures begin to fall.

At first, the change is gradual.

But as cooling continues, ice begins to form.

Ice reflects more sunlight than rock or water. This means more solar energy is sent back into space, less energy is absorbed at the surface, and temperatures drop further.

Cooling leads to ice, and ice leads to more cooling.

This is a feedback loop (see Figure 5.1).

Figure 5.1 — Climate Feedback Loop

Cooling–Ice–Albedo Feedback and CO_2-Driven Warming

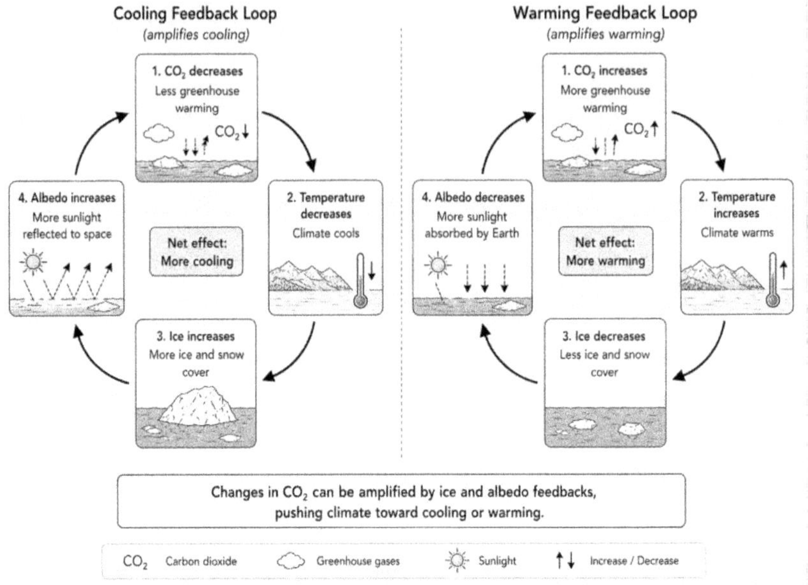

Once it begins, it can strengthen itself.

On a well-regulated planet, feedback loops like this are balanced by opposing processes. Volcanic activity, for example, releases carbon dioxide back into the atmosphere. Over time, that carbon dioxide can rebuild greenhouse warming and counteract cooling.

But the timing matters.

If chemical weathering removes carbon dioxide faster than volcanic activity replaces it, cooling can continue. Ice spreads. More sunlight is reflected away. The system moves farther from its previous state.

Eventually, as ice expands and covers more land, weathering slows. Exposed rock becomes less available. Rain has less direct access to fresh minerals. Volcanic carbon dioxide can then accumulate in the atmosphere over long periods.

This buildup can help the planet warm again.

But the return is not smooth.

The planet does not simply ease back into balance. It may swing from deep cold toward intense warming as carbon dioxide accumulates and ice retreats.

This pattern is often discussed in connection with the Snowball Earth hypothesis, in which much of the planet becomes covered in ice. Scientists continue to debate whether the oceans freeze completely or whether some open water remains. But the broader point is clear: Earth's climate system is capable of moving toward extreme states.

Rodinia does not maintain a narrow range of conditions.

It shifts.

These shifts have consequences for life.

Stable environments allow biological systems to develop and persist. When conditions change slowly and within limits, life can adapt. When conditions change rapidly or move beyond those limits, survival becomes more difficult.

In a system that overcorrects, stability is interrupted.

Periods of relative balance are followed by periods of disruption. Environments that support life may persist for a time, then change in ways that reduce their viability.

The oceans continue to provide protection, but they are not immune to these changes. Cooling affects ocean circulation. Ice cover alters the exchange of gases between the ocean and the atmosphere. Chemical conditions shift as temperature, circulation, and ice cover change.

Life remains present.

But it must endure variability.

The land, already constrained by radiation, limited water retention, and the absence of developed soils, becomes even less usable under these conditions. Cold periods increase the extent of ice. Warm periods do not immediately restore stability.

The system does not return smoothly to a balanced state. It swings.

Rodinia is a world where the components of climate regulation exist, but their interaction remains poorly tuned.

Processes that might stabilize the system can also amplify change.

The result is a planet that does not easily settle.

It responds to disturbance with further disturbance.

This pattern reinforces the constraints already in place.

Radiation limits surface exposure. Water limits where life can persist.

Climate instability disrupts continuity. Each factor alone creates difficulty; together, they define the system.

Rodinia is not simply a harsh environment.

It is a system that still cannot regulate itself effectively.

That instability reaches deeper than weather or ice. It is tied to the internal engine of the planet, where heat, tectonic motion, and volcanic activity keep reshaping the surface over time.

Chapter 6

A Faster, More Active Earth

The instability seen at the surface of Rodinia does not begin there.

It starts deeper.

The Earth during this time is not operating at the same pace as it does today. The interior contains more heat. That heat drives motion within the mantle, the thick layer of rock beneath the crust. Although the mantle is solid, it flows slowly over long periods of time.

When more heat is present, that flow can become more active.

Material rises in some regions and sinks in others (see Figure 6.1). This movement transfers energy from the deep interior toward the surface. It also helps move the rigid outer shell of the Earth, the tectonic plates.

During the time of Rodinia, this system may have been more energetic than it is today.

Plates move.

Boundaries shift.

The surface is continually reshaped.

This motion is slow on a human scale, but over millions of years it becomes significant. Continents collide, separate, and reorganize. Ocean basins open and close. Volcanic activity releases material from the interior into the atmosphere.

The system is in motion, but not steady.

Rotation adds another layer.

Figure 6.1 — Internal Heat Flow and Volcanism

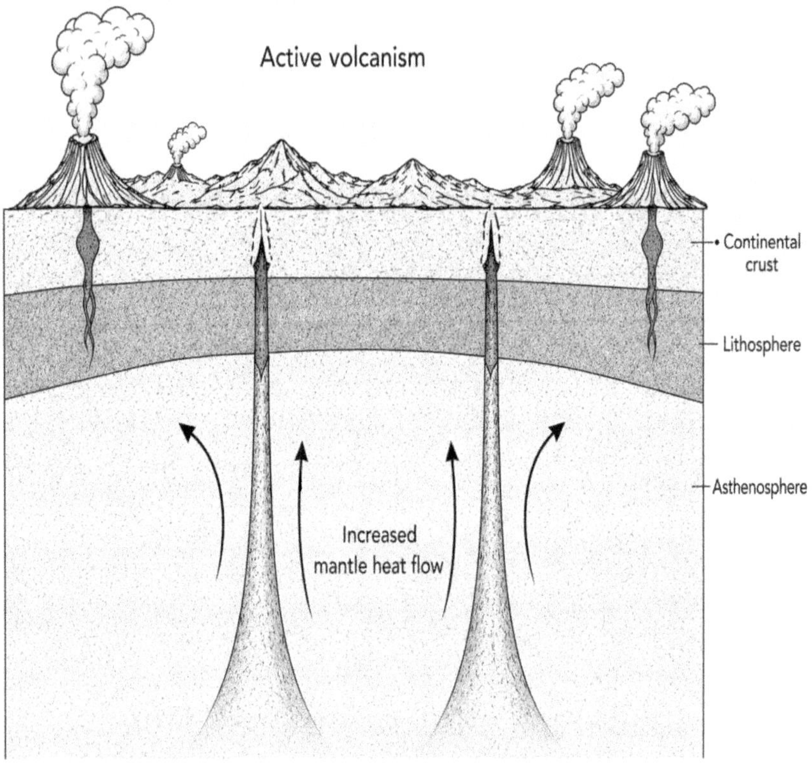

Internal heat drives mantle motion and volcanism, releasing gases that can influence climate and ocean chemistry.

Earth spins faster than it does today. This means the days are shorter.

The reason is that Earth has been slowing down over time. Much of that slowing comes from its relationship with the Moon. The Moon's gravity pulls on

Earth's oceans and creates tides. But because Earth is spinning, the tidal bulge does not sit perfectly under the Moon. It gets pulled slightly ahead by Earth's rotation. That offset creates drag. Over enormous spans of time, that drag acts like a very weak brake on Earth's spin. Earth slowly loses a little of its spin, and the Moon gradually moves farther away. As Earth's spin slows, the day becomes longer.

During the time of Rodinia, less of that slowing had occurred. The Moon was closer than it is today, and Earth turned through its daily cycle more quickly. Daylight and darkness followed one another at a faster rhythm.

That faster rhythm matters because rotation helps organize the movement of air and water. Winds, currents, storms, and temperature patterns all respond to how quickly the planet turns. A shorter day changes how heat builds during daylight, how quickly it is lost at night, and how energy is moved from one region to another.

The exact pattern would depend on many factors, including the atmosphere, the oceans, the strength of tides, and the arrangement of the continents. But the larger point is clear. Rodinia existed on an Earth with a different planetary rhythm: shorter days, a closer Moon, stronger tidal influence, and circulation patterns that may not have matched those of the modern world.

Climate is shaped not only by temperature, but by how heat is transported. Circulation patterns move heat across regions, influence rainfall, and help determine where conditions become warmer, colder, wetter, or drier.

When those patterns shift, the effects can be felt across the surface.

The result is a system that does not hold one pattern easily.

The interior and surface are connected.

Heat from the mantle drives tectonic activity. Tectonic activity influences volcanic output. Volcanic output affects atmospheric composition. Atmospheric composition influences temperature and climate.

Each part affects the others.

When the system is more active, these connections become more important.

Volcanic activity plays a key role.

When magma reaches the surface, it releases gases into the atmosphere, including carbon dioxide. Over long periods, this can increase the greenhouse effect and raise global temperatures. At the same time, major eruptions can send ash and sulfur-rich particles high into the atmosphere. These particles can reflect some sunlight back into space, causing temporary cooling.

These effects do not occur in isolation.

They interact with existing conditions.

Depending on timing and scale, volcanic activity can reinforce existing trends or push the system in a different direction. A long period of carbon dioxide release can contribute to warming. A major eruption can produce short-term cooling. The response depends on how the volcanic pulse fits into the larger climate system.

The system does not dampen change easily.

It transmits it.

This helps explain why Rodinia does not settle into a narrow, stable state. The forces driving change are strong, and the mechanisms that might limit that change are not always effective enough to keep conditions within tight boundaries.

Even the continents themselves contribute to this pattern.

As plates move, they alter the distribution of land and ocean. This affects how heat is absorbed and reflected, how moisture moves through the atmosphere, and where precipitation occurs. Large landmasses create different circulation patterns than fragmented continents.

Rodinia, as a large assembled landmass, interacts with these processes in a distinct way.

The scale of the continent influences the scale of the response.

A larger landmass can amplify extremes in temperature and moisture distribution. Interior regions can become drier. Coastal regions can behave differently from inland regions. Seasonal contrasts can become stronger.

Combined with an active internal system, this leads to greater variability across the surface.

The oceans continue to moderate conditions locally.

But at the planetary level, variability remains.

The key point is not that the Earth is chaotic.

It is that it is still adjusting.

The internal engine is strong.

The surface systems are developing.

The connections between them are active but not fully balanced.

Rodinia exists within this system.

It is shaped by forces that are powerful and persistent, but not coordinated in a way that produces long-term stability across the surface.

The result is a world where change is constant: not sudden, not random, but continuous and only partly controlled.

This ongoing activity reinforces the constraints already present at the surface. It limits the development of stable environments and makes it harder for conditions to remain consistent over long periods.

Those forces act through the shape of the world itself. Rodinia's size becomes part of the problem, because a supercontinent creates its own limits on climate, moisture, and habitability.

Chapter 7
The Supercontinent Problem

Rodinia is not just a landmass.

It is a very large landmass.

That scale changes how the system behaves.

On the modern Earth, continents are separated by oceans. These oceans act as reservoirs of heat and moisture. They absorb energy from the Sun, store it, and release it slowly. They also supply water to the atmosphere. Winds carry that moisture inland, where it falls as rain or snow and helps connect coastal regions with interior landscapes.

This moderates conditions.

Temperatures still vary.

Rainfall is still uneven.

But the oceans help reduce extremes and keep parts of the climate system linked.

Rodinia works differently.

When much of the continental crust is assembled into one large landmass, the distance from the ocean to the interior increases. Air moving inland loses moisture along the way. Rain falls closer to coastal margins, mountain belts, and other regions where air is lifted or cooled. By the time that air reaches the deeper interior, much of its moisture may already have been removed.

The result is dryness (see Figure 7.1).

Figure 7.1 — Supercontinent Interiors and Aridity

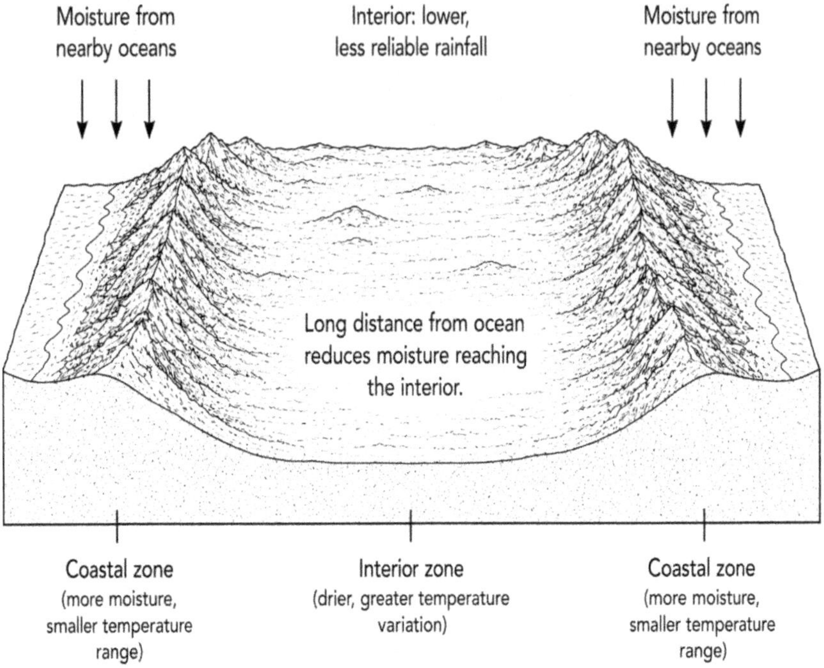

Large continental interiors tend to be drier and experience greater temperature extremes.

Not everywhere, and not always in the same way.

But as a broad pattern, the interior of a supercontinent is likely to receive less reliable moisture than its margins.

Water controls what the surface can support.

Without a steady supply of moisture, surface water becomes less dependable. Rivers are harder to sustain. Lakes, if present, may expand and shrink with changing conditions. In some regions, evaporation can exceed precipitation for long periods.

Water becomes a limiting factor, and temperature adds another constraint.

Water absorbs heat slowly and releases it slowly. Land responds more quickly. Near the coast, nearby oceans help soften temperature change. In the interior of a large continent, that moderating influence is reduced.

Heat builds more easily and is lost more quickly.

This creates wider temperature ranges between day and night and between seasons. Some regions can become very warm during periods of high solar input and much cooler when that input decreases.

Variability increases.

These conditions are not unique to Rodinia. They can be seen in smaller form on modern continents. Interior regions of large landmasses often experience lower precipitation and greater temperature variation than coastal regions.

Rodinia extends this pattern to a much larger scale.

The interior is not only far from the ocean.

It is partly cut off from the ocean's stabilizing influence.

This isolation has consequences.

Moisture does not reach all regions efficiently.

Temperature extremes become more pronounced.

Environmental stability is reduced.

Even if other conditions are more favorable, these factors still limit the usefulness of large portions of the land.

But Rodinia does not exist apart from the other constraints already described.

Atmospheric protection remains incomplete.

Surface water is not retained effectively because mature soil systems are absent.

Climate patterns are vulnerable to feedback processes that can amplify change.

The size of the supercontinent adds another layer.

It does not create an entirely new problem.

It enlarges the problems already present.

Dry conditions make it harder for water-dependent processes to persist. Temperature extremes increase stress on any system trying to stabilize. Isolation reduces the ability of different regions to balance one another through the movement of moisture, heat, and nutrients.

The land remains present.

But large portions of it are not only exposed.

They are also dry, variable, and distant from the moderating influence of the oceans.

The same pattern appears again.

The oceans support life.

The land limits it.

Rodinia, as a supercontinent, strengthens that division.

The edges of the landmass, where land meets water, remain the most active regions. These margins are more likely to receive moisture, exchange material with the oceans, and experience conditions that shift less severely than the deep interior. The interiors remain more constrained. Across the landmass, conditions vary, but few regions provide the combination of moisture, stability, and protection needed for sustained biological systems.

The problem is not a lack of space. It is a lack of usable space.

Rodinia offers vast areas of land, but much of that land does not meet the conditions needed for life to expand beyond the oceans.

That distinction is important.

It shows that the limitation is not simply geographic.

It is environmental.

Even where some moisture reaches the surface, another problem remains. Water alone is not enough. For land to become biologically useful, it must be held in place, nutrients must accumulate, and surfaces must develop structure. Rodinia's land largely lacks that foundation: mature soil systems capable of retaining water, storing nutrients, and supporting life over time.

Chapter 8
A Surface Without Foundation

Water reaches the land, but it does not stay there.

This is not only because of climate. It is also because of what is missing from the surface.

On the modern Earth, land is not just exposed rock. Much of it is covered by soil, a layered system made of mineral particles, organic material, water, air, and living organisms. Soil absorbs rainfall, stores moisture, and releases it slowly. It holds nutrients in place. It supports plants, which stabilize the surface, slow erosion, and add more organic material as they grow and decay.

Soil is not just a thin covering.

It is a system that allows land to function (see Figure 8.1).

On Rodinia, that system is largely absent.

There are no forests, grasses, shrubs, or root systems. There is no widespread organic layer built from leaves, stems, roots, and decaying land plants. There are no plant roots breaking rock into smaller particles, binding those particles together, and holding water near the surface. There are no complex soil ecosystems cycling nutrients through the land in the modern sense.

Figure 8.1 — With Soil vs. Without Mature Soil

Mature soil is the foundation of stable land systems.

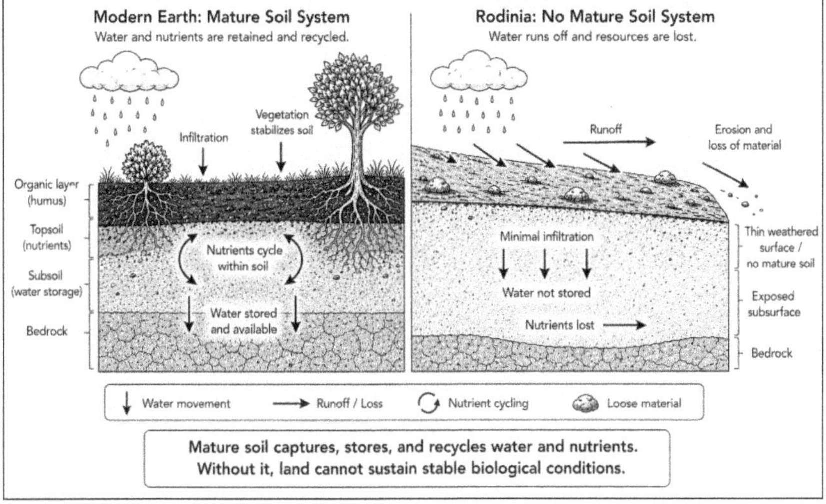

The land is mostly exposed rock, sediment, and weathered mineral surfaces.

When rain falls, it interacts directly with that surface. Some chemical weathering occurs. Minerals dissolve. Small particles are loosened. Thin films or temporary wet areas may form in some places. But without mature biological systems to retain and organize this material, much of it is easily removed.

Water moves across the surface.

It carries sediment with it.

It does not remain long enough, or consistently enough, to build stable structure across large areas.

This limits what can develop.

Without mature soil, water cannot be stored effectively. Without stored water, conditions fluctuate more rapidly. Wet periods are shorter. Dry periods dominate. There is no broad buffering system to smooth the transition between them.

Nutrients behave in a similar way.

On modern land, nutrients are held within soil and recycled through biological systems. Plants take up nutrients. Animals consume plants. Fungi, bacteria, and other decomposers return material to the soil. Over time, this cycle helps maintain availability.

On Rodinia, there is no comparable land-based cycle.

Nutrients released from rock are not held in place by roots, organic matter, and structured soil. They are more easily transported away by water. Some are carried into the oceans, where they can contribute to marine systems. On land, they are not retained in the same lasting way.

The surface cannot build complexity.

Even in regions where moisture is present for longer periods, the absence of mature soil limits development. Water may collect temporarily. Chemical reactions may occur. Sediment may accumulate in basins or low areas. But without a stable medium to hold water and nutrients together, conditions remain difficult to sustain.

Rain falls, reacts with rock, and moves on.

The land does not accumulate the materials needed to become biologically stable.

The earlier constraints deepen.

Radiation limits exposure at the surface.

Climate variability disrupts continuity.

The size of the supercontinent reduces moisture in the interior.

The absence of mature soil adds another layer.

It prevents the formation of stable, self-sustaining systems on land.

Life cannot expand far into an environment that cannot hold the resources it needs.

This is a structural limitation.

The land lacks the base layer required to support even simple land ecosystems across large areas. There is no widespread system on which more complex biological communities can build. Without that base, each event remains more isolated. Water arrives and departs. Nutrients are released and removed. Surface materials shift, erode, or wash away.

The result is a surface that exists physically but does not function biologically in the way later land surfaces will.

This does not mean the land has no influence.

As rain interacts with rock, it continues to break down minerals and transport them. These materials are carried into rivers, basins, shorelines, and oceans. There, they accumulate and become available to marine systems. In this way, the land contributes to life.

But it does so indirectly.

It feeds the oceans, but it does not support widespread life itself.

Rodinia, therefore, is not a balanced distribution of environments.

The oceans sustain life.

The land supplies material.

The connection between them is active, but uneven. Material moves from exposed land into water. Nutrients enter marine environments. The oceans become the place where those resources can be held, mixed, recycled, and used.

That uneven connection leads directly to the next part of the system: the movement of nutrients and chemical elements from exposed land into the marine environment, where they can support biological systems in ways the land cannot.

Chapter 9
Oceans Fed by the Land

The land does not support widespread life, but it still plays a role.

Every time rain falls on exposed rock, a process begins. Water interacts with minerals at the surface. Some minerals dissolve. Others break into smaller particles. This is chemical and physical weathering.

On Rodinia, this process is widespread.

There is no vegetation to slow runoff.

No mature soil to absorb and store water.

No root systems to stabilize the surface.

Water moves directly across rock, sediment, and weathered mineral surfaces.

As it moves, it carries material with it.

Much of this material does not remain on land.

It is transported.

Streams and rivers, even if temporary, move dissolved elements and fine particles toward lower elevations. Over time, some of these materials reach lakes, basins, shorelines, and oceans. There, they can accumulate, mix, and participate in ongoing chemical processes.

The oceans receive repeated input from the land.

From a land perspective, this is loss.

From an ocean perspective, it is supply.

This creates a clear contrast between the two environments.

The land releases material.

The oceans are better able to retain and process it.

This difference shapes where life can persist.

In the oceans, dissolved elements can remain available for longer periods. Nutrients circulate through water. Chemical reactions can occur repeatedly. Materials are mixed, redistributed, and reused. Systems can build over time because inputs are not immediately removed from the environment.

On land, the same materials are less stable.

They are released.

They are carried away.

They do not become part of a lasting surface system.

The ocean becomes the place where resources collect.

This is important for early life.

Microorganisms depend on access to basic chemical elements. These include carbon, nitrogen, phosphorus, sulfur, and trace metals. Some of these elements enter marine systems through weathering and runoff. Others are shaped by the chemistry of the atmosphere, volcanic activity, ocean circulation, and biological processes already occurring in the water.

Availability is not constant.

But in the oceans, it is more sustained.

In addition to supplying nutrients and dissolved ions, weathering also affects atmospheric conditions. Carbon dioxide in the atmosphere can dissolve into rainwater and react with exposed rock. Over time, some of that carbon is carried into the oceans as dissolved carbon compounds and can eventually become locked into minerals or become part of the ocean's ongoing chemistry.

This process can reduce the amount of carbon dioxide in the atmosphere.

As described earlier, that influences temperature.

The land, therefore, is connected to both ocean chemistry and climate.

It does not support widespread life directly, but it affects the environments that do.

This creates a system with uneven roles.

The land contributes material.

The oceans mix, retain, and process it.

Life develops where that processing can continue.

This relationship also introduces a form of dependency.

Marine systems benefit from continued input from the land. Without weathering, some nutrient supplies can decrease over time. With weathering, those supplies are renewed.

The balance is not precise.

Rates of input can change. Climate shifts can alter how much material is transported. Tectonic activity can expose new rock surfaces or bury older ones. Volcanic activity can add material from the interior. Ocean circulation can spread some elements widely and concentrate others in specific regions.

But the pattern remains.

The land feeds the oceans.

And in doing so, it reinforces where life is most able to persist.

This feeding is not controlled by land life.

It is driven mainly by physical and chemical processes.

Rainfall, temperature, rock composition, relief, and exposed surface area determine how much material is released. The absence of mature biological systems on land means there is little regulation from that side of the system.

The flow is mostly unmediated.

Material moves from land toward water with limited biological interruption.

The concentration of life in marine environments grows stronger.

Where resources accumulate and circulate, systems can develop. Where resources are removed quickly, systems struggle to persist.

Rodinia reflects this distinction clearly.

The land remains a source, the oceans remain the receiving environment, and life remains where the system allows continuity.

This continuity is not perfect.

Conditions in the oceans still change. Oxygen levels fluctuate. Temperature shifts occur. Chemical balances adjust over time. Nutrient availability rises and falls. Marine environments are not uniformly stable or equally habitable everywhere.

But compared to the land, the oceans provide a more reliable environment.

They hold what the land cannot.

Yet even the oceans do not achieve full stability, because their chemistry remains tied to an atmosphere still finding its balance, especially in the behavior of oxygen.

Chapter 10

Oxygen That Does Not Settle

Oxygen is present during the time of Rodinia, but it does not behave the way it does today.

On the modern Earth, oxygen exists at a relatively stable concentration in the atmosphere. It supports complex life and helps maintain a protective ozone layer in the upper atmosphere. These two roles are connected.

Oxygen supports respiration.

Ozone helps provide protection.

During the time of Rodinia, oxygen exists, but it is lower and less reliable than it is today. It is being produced and consumed at the same time, and the balance between those processes continues to shift.

Production occurs primarily in the oceans. Microorganisms use sunlight to drive chemical reactions that release oxygen as a byproduct. This oxygen enters the surrounding water and, under the right conditions, can eventually move into the atmosphere.

Consumption occurs through multiple pathways.

Oxygen reacts with dissolved iron and other reactive chemicals in the oceans. It reacts with gases released by volcanic activity. It reacts with exposed rock surfaces and weathered minerals on land. Each of these reactions removes oxygen from the system.

The result is variability.

Oxygen levels rise in some periods.

They fall or remain limited in others.

The system still does not produce the high, steady oxygen conditions familiar on the modern Earth (see Figure 10.1).

This instability has direct consequences.

Ozone forms from oxygen. When atmospheric oxygen is sufficient and sustained, ozone can accumulate in the upper atmosphere and provide a more reliable barrier against ultraviolet

Figure 10.1 — Oxygen and Ozone Instability

Oxygen is produced and consumed at the same time, so levels rise and fall rather than staying stable.

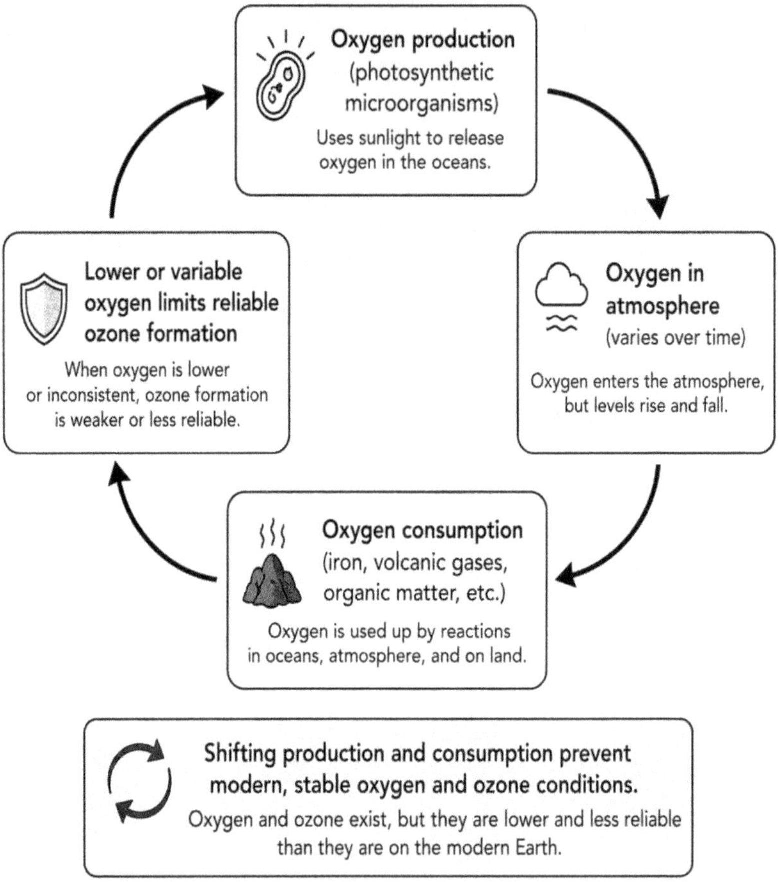

radiation. When oxygen levels are lower or more variable, ozone formation is also less reliable.

At times, the ozone layer may offer meaningful protection.

At other times, that protection may be thinner or less complete.

This repeats a pattern already seen across the system.

Protection exists.

But it does not hold in the modern sense.

The effects are not limited to radiation.

Oxygen also plays a role in chemical processes within the oceans. It helps determine how certain elements move through the water, how dead organic material is broken down, and how much energy living organisms can extract from their environment. When oxygen is more available, some reactions can proceed more efficiently. When oxygen is limited, those reactions slow down, shift, or happen in different ways.

This matters because oxygen is not just something organisms breathe. It is also part of the ocean's chemistry. It affects which nutrients remain available, how quickly waste material is recycled, and which environments can support more active forms of life. When oxygen levels rise or fall, the biological system changes with them.

This adds another layer of variability.

Marine environments remain more stable than land, but they are not fixed. Changes in oxygen availability affect where organisms can live and how they function. Some regions may support biological activity for extended periods, while others become less favorable.

Life adapts within limits.

But those limits remain narrow.

The instability of oxygen also helps explain why more complex systems remain limited. Higher oxygen availability supports more energy-intensive forms of life. When oxygen remains low, uneven, or inconsistent, those forms are constrained.

The problem is not only quantity.

It is also a problem of reliability.

Complex systems require conditions that persist. They require resources that are available not just occasionally, but consistently enough for biological systems to build on them. When key components fluctuate, development remains limited.

Oxygen during the time of Rodinia does not provide that reliability.

It is produced.

It supports limited processes.

It is consumed.

The balance shifts again.

This pattern connects back to other parts of the system.

Climate and oxygen are connected as well. Rainwater absorbs some carbon dioxide from the air and reacts with exposed rock, carrying some of that carbon toward the oceans. When carbon dioxide is removed from the atmosphere, temperature can change. Temperature affects ocean conditions. Ocean conditions influence biological activity, and biological activity affects oxygen production. Oxygen then reacts with materials in the ocean, atmosphere, and exposed land surfaces, reducing its concentration.

Each process feeds into the next.

None of them operates independently.

Rodinia is a system in which the necessary components are present, but their interactions do not produce modern stability.

Oxygen is part of that pattern.

It contributes to protection, but does not fully stabilize it.

It supports life, but within limits.

It participates in cycles, but does not dominate them.

The result is an environment where key conditions are intermittently or unevenly met, but not sustained in the way later complex life will require.

The broader constraint remains in place.

Life exists, but it remains limited in scope, location, and complexity.

Taken together, these limitations explain why the presence of land, water, energy, and oxygen is not enough. The next question is why those ingredients still do not produce widespread complexity.

Chapter 11

Why Complexity Does Not Happen

By this point, the pattern has become clear.

The components required for a complex world are present.

There is land.

There are oceans.

There is an atmosphere.

There is energy from the Sun.

Life already exists.

But widespread complexity does not develop.

The reason is not a single missing element (see Figure 11.1).

It is the interaction of multiple constraints.

Each system works to a degree, but none of them works consistently enough.

Start with protection.

Radiation levels remain elevated because atmospheric protection is incomplete. Oxygen exists, but it is lower and less reliable than it is today. Because ozone depends on oxygen, the ozone layer does not provide the same stable barrier found on the modern Earth. Other forms of shielding also help reduce exposure, but the surface still lacks a reliable protective system.

On land, this matters immediately.

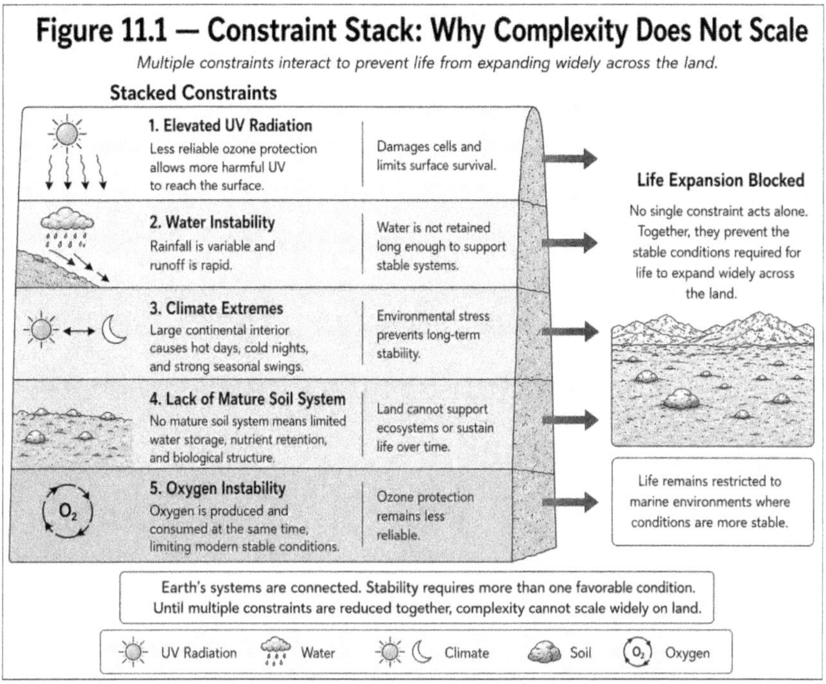

Figure 11.1 — Constraint Stack: Why Complexity Does Not Scale

Exposure is direct.

There is no buffer.

Water introduces the next constraint.

The oceans provide stability, but they also limit expansion. Life can persist there because water reduces radiation, moderates temperature, and holds dissolved chemistry in place. But this same environment confines life to zones where light, nutrients, and chemical conditions are suitable.

On land, water does not remain.

Without mature soil, rainfall moves quickly across the surface. There is no widespread system to store moisture or regulate its availability. Wet conditions are temporary. Dry conditions dominate.

This prevents continuity.

Climate adds another layer.

Rodinia does not regulate temperature efficiently. Processes that could stabilize the system can also amplify change. Cooling can lead to extensive ice formation. Warming can intensify atmospheric and oceanic shifts. These changes occur over long periods, but they disrupt the continuity needed for stable environments.

Conditions change.

They do not remain within narrow limits.

The size of the supercontinent increases these effects.

Large interior regions are likely to be drier and experience greater temperature variation. Moisture from the oceans does not reach far inland as reliably. The moderating influence of nearby water is reduced. Environmental conditions become more extreme.

Even where moisture is present, the surface cannot support it well.

Without mature soil, nutrients are not retained. Materials released from rock are transported away. There is no land-based cycle to capture, store, and reuse them. Each event remains more isolated.

Water arrives, reacts, and leaves. Little accumulates into a stable surface system.

Oxygen introduces another limitation.

It is present, but lower and less stable than it is today. It is produced largely in the oceans and removed through chemical reactions with reduced materials in water, rock, volcanic gases, and organic matter. Because of this, it does not maintain the kind of high, consistent atmospheric concentration that later supports complex life and a more reliable ozone layer.

Again, the pattern repeats.

A system exists.

But it does not hold in the modern sense.

Each of these constraints alone makes expansion more difficult.

Together, they prevent it.

Complex systems require continuity.

They depend on stable inputs of energy, water, and nutrients. They require protection from environmental extremes. They build on previous conditions and extend them over time. A complex system cannot develop if every supporting condition appears only temporarily, unevenly, or in the wrong place.

Rodinia does not provide that continuity.

It provides intermittent support.

Life persists in protected environments. It adapts within narrow ranges. It continues basic processes.

But it does not expand widely into new domains.

The limitation is stability.

The Earth system contains many of the ingredients needed for greater complexity, but those ingredients have not been coordinated into a stable whole.

Rodinia is not a world without life.

It is a world where life cannot scale.

This distinction is important.

It shifts the focus from absence to constraint.

The question is not why life is missing.

The question is why it cannot extend beyond its existing boundaries.

The answer lies in the interaction of systems that are present, active, and incomplete.

They reduce extremes, but not consistently. They support processes, but not continuously. They allow survival, but not wide expansion.

Rodinia represents a stage in which some parts of the Earth system are moving toward the conditions needed for greater complexity, while the whole system remains short of the coordination required.

Chapter 12
A World That Keeps Resetting

Rodinia does not simply hold life back.

It keeps changing the conditions under which life has to operate.

That distinction matters. The previous chapter described the limits themselves: incomplete protection, unstable oxygen, poor water retention, exposed land, uneven climate, and a supercontinent whose interior was difficult to sustain. This chapter follows what happens when those limits do not stay in one arrangement for long.

A coastline shifts.

A shallow sea expands, then retreats.

A region that receives moisture becomes drier.

Ocean circulation reorganizes.

Volcanic activity changes the atmosphere.

A setting that works for a time stops working in the same way.

This is the resetting problem.

Rodinia is not static. It is part of an active Earth system, and that activity continually changes the operating conditions at the surface. The changes are slow on a human scale, but over millions of years they are large enough to reshape environments entirely.

The source of this pattern lies beneath the surface.

Earth's interior remains active. Heat continues to move through the mantle, driving tectonic motion. Plates shift, collide, and separate. These processes reshape the crust over long periods, but their effects are continuous.

They do not pause.

As plates move, the arrangement of land and ocean changes. Coastlines migrate. Inland regions become more isolated or less isolated. Shallow seas appear where the crust lowers or where sea level advances, then disappear as conditions change again. Areas that once exchanged material with the ocean may become cut off. Other regions may open to circulation, moisture, and sediment.

Each change alters the setting in which life operates.

The change does not have to be sudden to matter. A shallow marine environment may remain favorable for a long period, but if circulation shifts, nutrients can become less available. If climate changes, temperature and oxygen conditions can shift. If land rises or sea level falls, the same region may become exposed, drier, or less protected.

Life can respond to change.

But response requires time.

That is the central difficulty. The environments most favorable to life are not always absent, but they are often temporary, shifting, or vulnerable to larger planetary changes. A region may provide sunlight, nutrients, water, and relative protection for a time. Then the system rearranges. The advantage weakens. Development is interrupted.

Volcanic activity adds another source of resetting.

When magma reaches the surface, it releases gases into the atmosphere. Carbon dioxide is among the most important. Over long periods, increased carbon dioxide can contribute to warming. At the same time, major eruptions can send ash and sulfur-rich particles high into the atmosphere. These particles can reflect some sunlight back into space and temporarily reduce incoming energy at the surface.

These effects do not balance each other in a controlled way.

They add variability.

Some processes push the system toward warming.

Others push it toward cooling.

The system adjusts, but not smoothly.

A warming interval can change ocean circulation, ice cover, rainfall, and the movement of nutrients from land into water. A cooling interval can restrict open water, alter gas exchange between the ocean and atmosphere, and reduce the area where biological activity can continue easily. Neither shift affects only one part of the system. Each one spreads through the connected machinery of land, ocean, atmosphere, and life.

The same pattern appears in the oceans.

Ocean environments are more stable than land, but they are not fixed. Water moves through circulation patterns that distribute heat, nutrients, oxygen, and other dissolved materials. When circulation changes, the conditions of marine life change with it.

Some regions become more favorable.

Others become less favorable.

A place that once received nutrients may receive fewer. A place that once held enough oxygen may become more limited. A region that once supported steady biological activity may become less reliable. The ocean remains the safest place for life, but even there, stability is conditional.

Rodinia keeps reconfiguring the map of opportunity.

The problem is not that every environment fails at once. The problem is that favorable conditions do not remain stable enough, broadly enough, or predictably enough. They appear in places. They persist for intervals. They allow life to continue.

Then the system changes again.

This does not erase life.

Life persists in protected environments, especially in the oceans. It adapts within the constraints it faces. It continues to carry out basic processes. But it does not build outward across the planet in the way later life will.

The system is active, dynamic, and continually reshaped.

That activity has two effects at the same time.

In the near term, it limits continuity. It interrupts favorable conditions before they can expand widely. It keeps environments from settling into long-lasting patterns.

Over the longer term, it prepares change. Tectonic motion exposes new rock, opens and closes basins, shifts coastlines, changes circulation, and redistributes heat and nutrients. Volcanic activity alters the atmosphere.

Weathering responds. Oceans reorganize. The same restless activity that prevents stability also creates new combinations of land, water, air, and chemistry.

Rodinia is therefore not just a constrained world.

It is a world being repeatedly reworked.

Each reset changes the system slightly. Some changes reinforce the old limitations. Others weaken them. New coastlines form. New marine settings appear. New circulation pathways become possible. The balance between atmospheric gases begins to shift. The shape of the supercontinent becomes less permanent than it once seemed.

This is why the resetting matters.

It does not merely repeat the same problem.

It gradually changes the structure of the problem.

For a long time, Rodinia holds the Earth system in a constrained state. But the forces beneath it do not leave that state untouched. They pull at it, heat it, fracture it, and reorganize it. The system does not yet become stable, but it begins moving toward a different arrangement.

The cycle begins to change when Rodinia itself starts to fragment, altering the balance among land, ocean, and atmosphere.

Chapter 13

The Break That Changes the System

Rodinia does not remain intact.

For millions of years, the Earth system keeps resetting. Coastlines shift, oceans reorganize, volcanic activity changes the atmosphere, and environments appear and disappear. But eventually, the resetting reaches the structure of Rodinia itself.

The same forces that assembled the supercontinent begin to pull it apart.

The process is slow, measured in millions of years, but its effects are large. The crust stretches, fractures form, and the landmass begins to separate and reorganize (see Figure 13.1).

This is not a surface event alone. It reflects changes deep within the Earth. Heat rising from the mantle weakens portions of the crust. As that crust thins, it becomes easier for it to break.

Where the crust separates, magma can rise.

This marks the beginning of rifting, the process in which a continent stretches, cracks, and begins to pull apart. Rifting does not split a continent instantly. It begins as zones of weakness in the crust. Over time, those zones widen, deepen, and allow molten material from below to move upward.

Figure 13.1 — From Rodinia to Pannotia

Continental breakup and reorganization create new coastlines, ocean basins, and changing circulation.

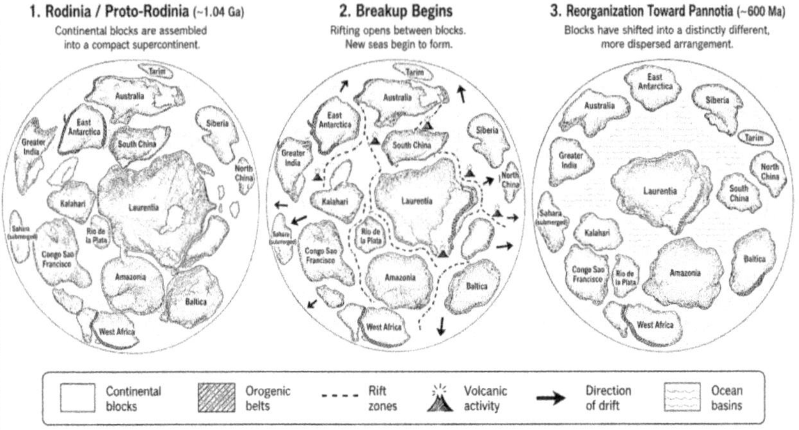

Rodinia's breakup opens new ocean basins and increases coastline length. Continued movement reorganizes the continental blocks into a later configuration often called Pannotia. Block positions are simplified and approximate.

As rifting continues, volcanic activity increases in some regions. Molten material reaches the surface. Gases are released into the atmosphere. Carbon dioxide is among the most important.

This changes the balance.

Earlier, chemical weathering removed carbon dioxide from the atmosphere and contributed to cooling. During rifting, volcanic activity can add carbon dioxide back into the system. At the same time, newly exposed rock can increase weathering, which can remove carbon dioxide. The result is not a simple switch from cooling to warming, but a changing balance between removal and addition.

The balance of the system begins to shift.

If carbon dioxide increases over long periods, the atmosphere can retain more heat. Extreme cold can begin to weaken. Ice that formed during earlier cooling periods can begin to retreat. Regions covered by ice may become exposed again. As ice cover decreases, the surface reflects less sunlight, and more solar energy can be absorbed.

This weakens the cooling feedback.

The system starts to move away from its most extreme cold conditions.

This does not happen instantly. It does not happen evenly everywhere. But over time, the breakup of Rodinia begins to alter some of the conditions that kept the system constrained.

At the same time, the breakup changes geography.

A single large landmass begins to divide into smaller pieces. As this happens, the distance between land and ocean decreases in many regions. Coastlines expand. Shallow marine environments become more common.

This matters because coastlines are zones of exchange. They are places where land, ocean, atmosphere, sunlight, minerals, and water interact more directly. Earlier chapters showed that the deep interior of Rodinia was difficult to sustain. It was farther from ocean moisture, more exposed to temperature extremes, and less connected to the stabilizing influence of water.

Breakup reduces that problem.

Moisture from the oceans can reach more areas of land.

Temperature variation becomes less extreme near new coastlines.

The moderating influence of water increases.

The scale of the supercontinent problem is reduced.

Conditions do not become stable immediately, but the structure of the world begins to change in a more favorable direction.

The oceans respond as well.

As new ocean basins form, circulation patterns change. Water can move through different routes. This affects how heat, nutrients, oxygen, and other dissolved materials are distributed. Regions that were once more isolated can become connected. Areas that had limited exchange can begin to interact with larger marine systems.

Connectivity increases.

Isolated regions begin to exchange water, heat, and chemistry.

This has implications for life.

More coastal areas mean more environments similar to the boundary zones described earlier. These areas provide access to sunlight, nutrients, water, and partial protection from radiation. They are not fully stable, and they are not yet modern ecosystems. But they offer more opportunities for persistence than the interior of a large continent.

The system opens.

This does not eliminate the older constraints.

Radiation remains a factor.

Oxygen is still variable.

Mature soil systems are still absent.

Climate remains uneven.

But the arrangement of the system is different. The land is no longer dominated in the same way by one vast interior. The oceans are no longer organized around the same boundaries. Coastlines multiply. Basins open. Circulation changes. The exchange among land, water, and atmosphere becomes more active.

Rodinia's breakup and later reorganization create a different kind of world.

The key change is not a single improvement.

It is the weakening of several constraints at once.

Where earlier limitations reinforced one another, the breakup begins to loosen that alignment. Conditions that once combined to prevent expansion no longer fit together in quite the same way.

This creates space within the system.

Not empty space.

Functional space.

Environments where processes can persist longer.

Regions where conditions may change more gradually.

Opportunities for systems to begin extending beyond their previous limits.

Over time, the moving continental blocks do not simply disappear into isolation. Some later come together in a different arrangement, often called

Pannotia. This later configuration is shorter-lived and more debated than Rodinia, but it reflects the same larger process: Earth's continents are not fixed. They assemble, separate, and assemble again in new forms.

That larger process matters more than the name of any one arrangement.

Rodinia moves from a state defined by constraint toward one where those constraints begin to loosen. The Earth system becomes less dominated by one vast interior and more shaped by coastlines, basins, circulation, and exchange.

The system is not stable.

But it is becoming more capable of supporting stability.

As geography opens, the atmosphere becomes part of the next question.

Oxygen and ozone do not matter only as isolated features. They begin to matter as parts of a protective system that can, over time, become more dependable.

Chapter 14

Protection Begins to Hold

The breakup of Rodinia changes more than the shape of the land.

It changes the conditions under which protection can become more reliable.

Earlier, protection exists but does not persist in the modern sense. Oxygen is present, but it is lower and uneven. The ozone layer forms, but its protection is not steady. Radiation reaches the surface in ways that limit where life can survive. The magnetic field provides background protection for the planet, but the atmosphere itself does not yet offer the steady shield familiar on the modern Earth.

That pattern begins to change.

Not all at once.

Not everywhere.

But enough for the direction of the system to shift.

As Rodinia fragments, the planet becomes less dominated by one vast interior and more shaped by coastlines, shallow seas, open water, and exchange between ocean and atmosphere. This does not create stability immediately. It does not remove every constraint. But it gives the system more ways to moderate itself.

That change matters for oxygen.

Oxygen is produced mainly by microorganisms in the oceans. These organisms use sunlight to drive chemical reactions and release oxygen as a byproduct. When open water is more available, and when shallow marine environments become more widespread, microorganisms that produce oxygen have more places to live and can remain active for longer periods.

The oceans are still variable.

But they are less confined by the old arrangement.

More coastlines mean more shallow water.

More shallow water means more places where sunlight, nutrients, and marine life can interact.

More interaction means more opportunity for oxygen production to continue.

Oxygen is still being consumed at the same time. It reacts with dissolved metals, volcanic gases, exposed rock, and other materials that remove it from the system. That does not stop. But the balance begins to change. Production can become more sustained in some environments. Fluctuation can begin to decrease.

The atmosphere does not suddenly become modern.

But it begins moving in that direction.

As oxygen becomes more consistent, ozone formation can also become more reliable. Ozone forms in the upper atmosphere from oxygen molecules. When oxygen is present in sufficient and sustained amounts, ozone can build into a layer that absorbs ultraviolet radiation before it reaches the surface.

That matters because ultraviolet radiation is one of the barriers that kept the land exposed.

Earlier, protection was partial. At times, the ozone layer may have reduced radiation significantly. At other times, lower or more variable oxygen would have made that protection less complete. The surface could not be depended on in the same way it can be today.

With more reliable ozone formation, that begins to change.

Radiation does not disappear.

But its effect becomes less severe.

Protection becomes more repeatable.

This is an important shift. Life does not require perfect protection. It requires protection that holds often enough and long enough for biological processes to continue. A surface that is exposed unpredictably is difficult to use. A surface where exposure is reduced more consistently becomes less hostile.

The difference is not absolute safety.

The difference is reliability.

The magnetic field remains part of the background protection of the planet, as described earlier. But the major change in this stage involves the atmosphere itself. As oxygen becomes more consistent, ozone can become more reliable, and ultraviolet radiation at the surface can become less severe. Water continues to provide protection in the oceans and shallow marine environments, but the atmosphere begins to carry more of the burden.

The system is not yet fully coordinated.

But one major limitation begins to weaken.

Radiation becomes less dominant.

That change has direct consequences for where life can persist. Lower radiation stress reduces pressure on biological systems. Environments that were too exposed begin to move closer to tolerable conditions. Shallow water regions, especially near expanding coastlines, experience less severe variation in exposure. The boundary between ocean and land becomes more accessible.

This does not mean life moves widely onto land.

It means the barrier is reduced.

The edge becomes more workable.

The same coastal and shallow marine environments opened by Rodinia's breakup now matter in a new way. They are not only places where water, minerals, and sunlight meet. They are places where protection is becoming more dependable. Organisms that remain tied to water can persist in settings closer to exposure than before. Biological activity can continue under conditions that would have been more difficult in the earlier Rodinian system.

This does not make the land immediately usable.

Other constraints remain.

Water is still not retained effectively on land.

Mature soil systems have not developed.

Climate variability still exists.

Oxygen is still not as high or stable as it will be later.

But the protective system is changing. It moves from stronger fluctuation toward greater stability. From partial protection toward more consistent protection. From variation toward repetition.

That repetition matters because biological processes depend on conditions that can recur. A process that works only briefly remains limited. A process that can repeat under similar conditions can begin to leave a larger effect.

More reliable oxygen and ozone do not complete the transition.

But they make repetition easier.

They allow protection to begin holding.

The change is gradual and uneven. Some regions remain harsh. Some environments remain unstable. The land is still largely without mature soil, deep biological structure, or reliable water retention. But the broader direction has shifted. Protection is no longer only present in scattered or temporary ways. It begins to become part of a more dependable system.

This transition does not complete the process.

It enables the next stage.

With protection becoming more reliable, the larger question changes. The issue is no longer only whether one system can improve. The issue is whether several improving systems can begin to support one another at the same time.

That is where possibility begins.

Chapter 15

From Constraint to Possibility

The world shaped by Rodinia does not become stable all at once.

But it begins to allow stability.

That distinction matters. Earlier in the story, the necessary pieces were already present. There was land, water, atmosphere, energy from the Sun, and life. But those pieces did not work together well enough to support widespread complexity. Land existed, but it could not hold water or nutrients. Oceans supported life, but mostly within limited zones.

Oxygen existed, but it was lower and uneven. Protection existed, but it did not hold consistently. Climate could shift sharply. Tectonic activity moved material and energy through the planet, but it also kept resetting environments before they could persist.

The problem was not absence.

It was coordination.

By this stage, that begins to change.

No single system becomes modern. No single improvement solves the problem. The land does not suddenly become fertile. The oceans do not become uniformly stable. The atmosphere does not become identical to the atmosphere of today. Life does not immediately spread across the continents.

But several conditions begin to improve together.

That is the important shift.

Protection becomes more dependable.

Coastlines expand.

Shallow marine environments become more common.

Ocean circulation changes.

Moisture can reach more regions.

Temperature extremes begin to moderate in some settings.

Nutrients can move through a more connected system.

Biological activity can continue for longer intervals.

Each change matters by itself.

Together, they matter more.

A planet becomes more habitable not only when one condition improves, but when several conditions begin to reinforce one another. More open water supports biological activity. Biological activity can contribute to oxygen production. More consistent oxygen can support more reliable ozone formation. More reliable ozone reduces radiation stress. Reduced radiation stress makes shallow and marginal environments less hostile. More coastlines create more places where those conditions can interact.

The result is not full stability.

It is the beginning of support.

Earlier, the constraints worked together in the opposite direction. Large interiors reduced moisture. Poor water retention limited land surfaces. Incomplete protection increased exposure. Climate instability disrupted continuity. Oxygen variability limited biological activity. Each constraint made the others harder to overcome.

Now that alignment begins to loosen.

The system does not simply stop being difficult.

It becomes less locked into difficulty.

This is the difference between a world that blocks expansion and a world that begins to permit it.

The breakup of Rodinia is central to that change because geography controls many other parts of the system. When one large landmass begins to divide, the relationship between land and ocean changes. The distance from the sea decreases in many areas. Coastlines multiply. Shallow seas and marginal environments become more common. Ocean basins open and circulation patterns reorganize.

This creates more exchange.

Land and ocean interact across longer boundaries. Water carries dissolved minerals and nutrients into marine environments. Oceans redistribute heat and chemistry. Coastal regions experience less extreme conditions than deep interiors. Shallow water environments receive sunlight while still retaining some of the protection and chemical continuity of the ocean.

These are not minor changes.

They change where persistence is possible.

Persistence is the key.

Complexity does not develop from ingredients alone. It develops when processes can continue long enough for one change to build on another. A favorable environment that appears briefly may support survival, but not expansion. A favorable environment that persists can support accumulation. Biological activity can repeat. Chemical cycles can continue. Oxygen production can become more sustained. Nutrients can remain available for longer periods. Small advantages can begin to matter.

Rodinia itself does not produce the full flowering of complex life.

But it helps change the conditions under which such development becomes possible.

The oceans remain the main stage. Life is still mostly aquatic. The land remains limited. Mature soil systems have not formed. There are no forests, grasses, or land ecosystems in the modern sense. Rain still runs across exposed surfaces more easily than it is retained. Nutrients are still more likely to be carried away than stored in place.

But the boundary between land and water becomes more important.

The edges of the system become more active.

At those edges, several improving conditions meet. There is water, sunlight, mineral input, partial protection, and increasing exchange with surrounding environments. These zones are still unstable, but they are less isolated than before. They provide more opportunity for biological processes to continue, adapt, and repeat.

This is where possibility begins to replace simple constraint.

Not because the world has become easy.

Because the world has become less closed.

Earlier chapters described a planet where the necessary components were present but misaligned. By this point, that misalignment has begun to change.

Protection becomes more reliable in some intervals. Open water becomes more available. Circulation becomes more connected. Coastlines create more places of exchange. Climate remains variable, but some environments experience less severe extremes.

The system begins to hold in pieces.

Those pieces matter.

A stable ocean region here.

A productive shallow sea there.

A coastline where materials accumulate.

A period when oxygen production continues longer than before.

A region where radiation stress is reduced enough for biological activity to persist closer to exposure.

None of these alone transforms the planet.

Together, they change the direction of the story.

This is the movement from constraint to possibility.

Possibility does not mean inevitability. It does not mean that complex life must appear immediately or that land must suddenly become habitable. It means the system begins to provide more openings than before. More places where conditions last. More ways for processes to connect. More chances for biological activity to extend beyond narrow limits.

The planet is still unfinished.

But it is becoming more workable.

Rodinia's importance lies partly in this transition. Its assembly creates large-scale constraints. Its instability reveals the limits of a poorly coordinated system. Its breakup changes the structure of the planet and opens pathways that were less available before.

That is why this stage matters.

The ingredients were already there.

Land.

Water.

Atmosphere.

Energy.

Life.

But now those ingredients begin to interact in more useful ways. They do not yet form the fully coordinated Earth system that later supports widespread complexity. But they begin moving toward it.

Protection starts to hold.

Water and coastlines create more exchange.

Oceans remain the primary refuge, but they become more connected and more productive.

Climate remains variable, but not every environment is reset in the same way at the same time.

Biological activity remains limited, but it has more opportunities to persist.

This is not the end of constraint.

It is the beginning of a different balance.

The world is still unstable, but it is no longer only a world of interruption. Some processes can continue. Some environments can last. Some connections can strengthen.

That is enough to change what comes next.

Rodinia does not complete the transition to a fully habitable land world. It does not produce the forests, soils, animals, and ecosystems that later transform the surface. But it helps move Earth toward the conditions that make those later transformations possible.

The system has not finally settled.

But it has begun to align.

Chapter 16
The Mother of Supercontinents

Rodinia does not stand alone.

It sits between worlds.

Earlier continental assemblies brought large portions of Earth's crust together before Rodinia. Later supercontinents, including Pangea, formed under conditions that allowed complex life to spread widely across land. Rodinia belongs between those states. It is not the beginning of continents, and it is not the world of forests, animals, soils, and fully developed land ecosystems.

Its importance lies somewhere else.

Rodinia helps mark a stage in which Earth's systems begin moving from presence toward alignment.

That distinction has shaped the entire story.

The ingredients were present before the world was fully workable. Land existed, but it did not yet function as a stable biological surface. Oceans existed, but life remained concentrated within them. Oxygen existed, but it was lower and uneven. Protection existed, but it did not hold consistently. Climate regulation existed, but it could overcorrect. Tectonic activity moved the planet forward, but it also kept resetting the conditions under which life had to operate.

Rodinia gathers those tensions into one planetary system.

It is a supercontinent, but it is more than a map of joined landmasses. Its size changes climate. Its interior becomes difficult to sustain. Its exposed surfaces feed the oceans but do not support widespread life. Its edges become places of exchange. Its assembly creates constraint, and its breakup begins to loosen that constraint.

That is why Rodinia matters.

It shows a planet that almost works.

The land is there.

The oceans are there.

The atmosphere is there.

Life is there.

But the systems do not yet reinforce one another reliably enough. They operate, but not with modern coordination. They support survival, but not broad expansion. They allow persistence in protected environments, but not the widespread development of complex ecosystems across the surface.

Rodinia's role is not to complete the transition.

Its role is to help reorganize the conditions under which the transition can occur.

The assembly of Rodinia creates large-scale problems. One vast landmass increases the distance between interior regions and the ocean. Moisture becomes less reliable inland. Temperature variation becomes more pronounced. Land remains exposed, dry, and poorly buffered. Without mature soil systems, water and nutrients are not held in place. The surface remains physically present but biologically limited.

At the same time, Rodinia connects systems that had to interact.

Weathering on exposed land affects carbon dioxide. Carbon dioxide affects temperature. Temperature affects ice, ocean circulation, and biological activity. Biological activity affects oxygen. Oxygen affects ozone. Ozone affects radiation at the surface. Each process is linked to the others, even when the links do not yet produce stability.

This is the larger pattern.

Earth does not become habitable through one improvement.

It becomes habitable as multiple systems begin to support one another.

Rodinia is important because it sits within that movement.

Its existence reveals the limits of a poorly coordinated world. Its breakup begins to change the structure of that world. Coastlines expand. Ocean basins open. Circulation patterns shift. Moisture can reach more areas. Shallow marine environments become more common. Exchange among land, ocean, atmosphere, and life becomes more active.

None of this produces a modern Earth immediately.

But it changes what is possible.

That is the difference between an event and a transition.

An event happens and passes. A transition changes the conditions that follow. Rodinia's assembly, constraint, breakup, and reorganization are part of such a transition. They do not explain everything that comes later, but they help move Earth away from a state where systems interfere with one another and toward one where they can begin to cooperate.

This is why the phrase “Mother of Supercontinents” matters.

It should not mean that Rodinia is the first supercontinent, or the only important one, or the single cause of later habitability. Its meaning is broader and more suggestive. Rodinia is a mother in the sense that it prepares, carries, and reorganizes conditions. It gathers earlier planetary instability into a form that can be broken, redistributed, and reshaped into something more workable.

It gives birth not to a finished world, but to a new direction.

The later Earth does not emerge from Rodinia fully formed. It emerges through continued change. Continents separate and assemble again. Atmospheres shift. Oceans circulate through new pathways. Oxygen becomes more dependable. Protection improves. Climate becomes more capable of supporting persistence in some environments. Eventually, much later, land surfaces develop the biological foundation they once lacked.

Soils form.

Water remains.

Nutrients are held and recycled.

Life expands beyond the oceans.

But that later story depends on a long preparation.

Rodinia is part of that preparation.

Its significance is not in hosting the final result. It does not contain the forests of the Carboniferous world, the animals of later seas, the soils that transform continents, or the ecosystems that eventually cover land. It belongs to an earlier and less complete Earth.

But incomplete does not mean unimportant.

Rodinia is a world where the limits are visible.

It shows why land alone is not enough.

It shows why oxygen alone is not enough.

It shows why oceans alone are not enough.

It shows why life alone is not enough.

The pieces must interact in ways that hold.

That is the central lesson of Rodinia.

A planet can possess the necessary ingredients and still remain constrained. It can have land without usable land, oxygen without stable protection, water without lasting soil, and life without broad expansion. The difference between a world that contains life and a world that supports complexity is not simply what exists. It is how long those conditions persist, how widely they apply, and how well they reinforce one another.

Rodinia stands at the point where that reinforcement begins to become more possible.

Its assembly exposes the problem.

Its breakup changes the problem.

Its legacy is the movement from interference toward coordination.

Earlier, constraints acted together. Radiation, climate instability, poor water retention, limited oxygen, exposed land, and supercontinent geography reinforced one another. After Rodinia begins to break apart, that alignment weakens. The system remains unstable, but it is less trapped in the same configuration.

Openings appear.

Coastlines multiply.

Oceans connect.

Protection improves.

Some environments last longer.

These changes do not complete the transition, but they matter because continuity is the foundation of development. Where conditions persist, processes can repeat. Where processes repeat, effects can accumulate. Where effects accumulate, change can move beyond survival and begin to produce new possibilities.

Rodinia helps create those possibilities.

Not directly.

Not completely.

But meaningfully.

That is its place in Earth's long story.

It is not the first world.

It is not the final world.

It is not the most familiar world.

It is one of the worlds that prepares the system for what follows.

The land that cannot yet be used begins to move toward usability. The protection that does not fully hold begins to become more reliable. The oceans that shelter life begin to interact with a more open geography. The atmosphere begins to participate in a more dependable protective system. The conditions that fluctuate begin, in some places and over time, to persist.

Rodinia does not produce the final outcome.

It helps make the outcome possible.

That is what defines it.

A system that does not fully work, but one being reorganized into something that can.

A world that holds the problem and begins to loosen it.

A supercontinent that helps prepare the Earth for everything that follows.

Epilogue
When the System Finally Holds

For most of Earth's history, the pieces were present.

Land.

Water.

Atmosphere.

Energy.

Life.

But presence was never enough.

The deeper question was whether those pieces could work together.

Rodinia belongs to the long middle of that question. It is not the beginning of Earth's story, and it is not the finished world that later supports complex life across land and sea. It is a world in between, a world where the necessary systems exist but do not yet hold together reliably.

Protection forms, but remains incomplete.

Climate responds, but often swings too far.

Water supports life, but mostly within limits.

Land exists, but cannot yet function as a stable living surface.

Oxygen is present, but uneven.

Again and again, the pattern appears.

The ingredients are there.

The system is not yet steady.

That is what makes Rodinia important.

It shows a planet that almost works.

The land stretches across the surface, but much of it cannot be used. The oceans shelter life, but life remains mostly confined. The atmosphere offers protection, but not with modern reliability. The climate changes, adjusts, and overcorrects. Conditions appear, shift, weaken, and return in altered form.

Life persists.

But it does not yet expand widely.

The difference is not life versus no life.

The difference is whether life can build outward.

Rodinia helps show why that difference matters. A planet can contain the right ingredients and still remain constrained. It can have land without usable land, oxygen without steady protection, water without lasting retention, and life without broad expansion.

The pieces must do more than exist.

They must reinforce one another.

Over time, that begins to change. Rodinia breaks apart. Coastlines expand. Oceans connect in new ways. Exchange increases among land, water, atmosphere, and life. Protection becomes more dependable. Some environments last longer than before. Some processes repeat often enough to matter.

This does not create the modern Earth.

It does not suddenly produce forests, soils, animals, or stable land ecosystems.

But it changes the direction of the system.

That is Rodinia's quiet importance.

It does not complete the transition.

It helps make transition possible.

The Earth we recognize comes later. A world where land can eventually be used. Where protection becomes more consistent. Where water and nutrients can be held. Where life is no longer confined to narrow refuges.

Rodinia is not that world.

It is one of the worlds that makes that world possible.

Its importance lies not in completion, but in preparation. Its assembly concentrates the problem. Its instability exposes the limits. Its breakup loosens the constraints. Its reorganization helps move Earth toward continuity.

The change is quiet.

It does not appear as a single boundary.

It appears as persistence.

A process lasts longer than before.

An environment holds together for more time.

A protective layer becomes more dependable.

A coastline supports more exchange.

A biological cycle repeats often enough to matter.

And once persistence becomes possible, everything that follows has something to build upon.

That is Rodinia's place in Earth's story.

Not the first world.

Not the final world.

Not the world where complexity fully arrives.

But the world where the difference begins to take shape.

The difference between having the pieces and having a system.

The difference between survival and expansion.

The difference between a planet that almost works and one that finally can.

www.ingramcontent.com/pod-product-compliance
Lightning Source LLC
LaVergne TN
LVHW050820061126
851857LV00012B/894
9798349514357